A Personal Search for Understanding Evolution:

Sharing the same world and similar genome... but not the same ancestry or destiny

BY

Jay D. Clark, M.D.

Dorrance Publishing Co
585 Alpha Drive
Pittsburgh, PA 15238
Visit our website at *www.dorrancebookstore.com*

ISBN: 978-1-6853-7195-1
eISBN: 978-1-6853-7737-3

Dedicated to my son, Kyle,
whose thought-provoking observations reopened
my search and motivated me to share it with others.

Table of Contents

Why Am I Writing About Evolution?

One of the most perplexing statements I hear people make is: "I lost my faith because of science." When I ask, "What science?" I am never given a clear answer as to how it was science's fault. The expression "one man's honey is another man's poison" (what is life to one may be death to another) aptly describes some people's reactions to faith and science. Although we may not be able to choose our allergies, we can turn our faith and science poisons into honey with the proper understanding.

My early military school education[1] was patterned after Gracian's Manual which states: "To jog the understanding is a greater feat than to jog the memory: for it takes more to make a man think, than to make him remember."[2] It was there that I was introduced to Darwinism which prompted me to study the subject on my own, weigh the evidence, and draw my own conclusions "on the origin of species."[3]

Even though my parents were active Latter-day Saints, they seemed pleased that I was being exposed to a theory that had received so much attention. My father was twenty years old when the Scopes Trial[4] of 1925 gained public notoriety. He was a devoted Baptist at the time who felt that the exposure of truth required both sides to present their best case. Dad loved what Voltaire's biographer said: "I may disapprove of what you say, but I will defend to the death your right to say it."[5]

At California Military Academy I became interested in the scientific method of observable, repeatable experimentation. I later learned that the gospel of Jesus Christ was observable, repeatable self experimentation.[6] Although my church, the Church of Jesus Christ of Latter-day Saints, did not have an official position on evolution, I recognized my own personal bias against it which I resolved to restrain so it wouldn't affect my pursuit of scientific truth. Science taught me that there was no such thing as a scientific lie. The opposite of scientific truth is myth, which I wasn't willing to settle for despite my most cherished beliefs.

Any personal guilt I felt over faith bias early on was later offset by the anti-faith bias I encountered in the scientific community. In academia, evolution is the *sacred shrine.* If one opposes it even with the soundest scientific data, that person risks opposition, anger, disdain, and expulsion from the academic world.[7] Richard Dawkins provides a reason for anti-faith bias: "Before Darwin it was impossible to be an intellectually fulfilled atheist."[8] Richard Lewontin adds, "We take the side of science in spite of the patent absurdity of some of its constructs because we have a prior commitment to materialism. Moreover, that materialism is absolute, for we cannot allow a Divine Foot in the door."[9]

These statements are not only anti-faith. They are also anti-science and suggest that evolution has difficulty defending itself against tough questions. In fact, they transform evolution from science to philosophy. For a time, this discovery caused me to undergo a shaken-science crisis akin to the shaken-faith crisis some people have when they face unmet faith expectations. Today, casual comments about evolution are frequently made without pushback. For example, a *Scientific American* writer who was pondering the question of how people differ from other primates wrote: "Humans have evolved a sense of self that is unparalleled in its complexity."[10] The author could have said, "Humans have a sense of self that is unparalleled in its complexity." There have been no studies demonstrating how evolutionary processes could produce a mind with a sense of self. The word "evolved" carries no information in this context. It just sounds science-y. Planting Darwin's flag everywhere without evidence is confusing to those trying to follow good standards of scientific investigation.

The scientific method cannot be strictly applied to the study of unobservable, unrepeatable past evolutionary history. However, a great deal can be learned about it from genetics and origin science which encompasses causality (everything that has a beginning has a cause) and analogy (intelligence is needed to generate complex coded information in the present, so we can reasonably assume the same for the past).

Darwin taught that all living things are modified descendants of a common ancestor by natural selection and that over time, evolutionary change gives rise to new species. During the 1930s and '40s, his theory was modified by Neo-Darwinism which explicitly denies any purpose, prediction, or programming in evolution. Engineer Perry Marshall emphasizes this modification by saying, "Because randomness and absence of purpose are essential to the Neo-Darwinian Modern Synthesis, I do not consider biologists who reject the 'random mutation' hypothesis to be Neo-Darwinists."[11] Over the years, Neo-Darwinism became the prevailing paradigm in science. However, ever since the discovery of the role of DNA in life, the number of evolution-rejecting scientists has been increasing. Neo-Darwinian evolution is more widely questioned today than at any time.[12]

WHAT IS EVOLUTION?

The word "evolution" comes from the Latin "evolutio" which refers to the purposeful act of unrolling, as in the unrolling of an ancient book. In English the word came to mean an unrolling or unfolding of that which already existed in a more compact form. Eighteenth-century biologists adopted the term to describe embryological development which conformed to their theory of preexisting miniatures. When epigenesis (creation of new structures or entities) was later discovered, the term evolution continued to be used. Implicit in its definition was the fact that embryology was not a random process. Some force (mechanical or vital) controlled development, ensuring that correctly shaped organs were produced within a short gestational period in the proper sequence. Non-scientific authors from the fields of language, politics and social organizations used "evolution" in a figurative sense to refer to almost any kind of connected sequence of events unfolding from a preordained design.

In his 1859 work, *On the Origin of Species*, Darwin did not use the term evolution. He wrote about transmutation (adaptation by natural selection) because evolution at the time meant either the unfolding of a predetermined plan, which Darwin wanted to put into question, or the gradual development of a system toward something better. The word "transmutation" simply meant change or transformation without implication of upward, horizontal or downward direction. Although Darwin believed transmutation would produce progression in the long run, progression was not part of his "theory of transmutation." It was nearly another decade before the title "theory of evolution" began to gain popularity. According to biological historian Peter J. Bowler, nineteenth-century philosopher Herbert Spencer "adopted 'evolution' as the general name for the process of development which he tried to trace out in every field from cosmology to the human mind, thereby laying the foundations of the modern usage of the term."[13]

Direction of change is critical in distinguishing evolution from non-evolution. For example, when the Latin word "vita" (life) was translated "vida" in Spanish and "vie" in French, no new information or improvement was conveyed within these cultural adaptations. Language evolution did not occur—only language change (transmutation). In any context, the meaning of evolution isn't simply the recycling of limited alternatives but must include planned growth and development. Unfortunately, as Professor Luc Steels states, "the idea that evolution implies change toward a better system is not at all implied by the contemporary notion of biological evolution."[14] Interestingly, Darwin's 1871 book was entitled *Descent of Man*.

What Is Life?

In the 1997 movie *Contact*, scientists from the Search for Extra-Terrestrial Intelligence (SETI) Agency scan space for unmistakable signs of intelligent life. What constitutes an unmistakable sign of intelligent life? A message. Jodie Foster gets excited when her antenna picks up radio waves that appear to have an intelligent pattern: "One, two, three, five, seven, eleven ... those are primes! That can't be natural phenomena!" One official asks with a hint of derision, "If this is such an intelligent source, then why don't they just speak English?" Foster fires back, "Because math is the only universal language!" *Contact* was based on a novel by the late Carl Sagan, an evolutionist who believed in spontaneous generation and who was instrumental in starting the real SETI program. To Dr. Sagan's observation, if a simple string of prime numbers proves the existence of an intelligent being, wouldn't the equivalent of a thousand encyclopedias in the first living cell be much more convincing evidence of an intelligent being?[15]

Functional information comes from intelligent minds. Scientists recognize the connection between information and intelligence. "DNA," according to the founder of Microsoft, Bill Gates, "is like a computer program but far, far more advanced than any software ever created."[16] If it's absurd to think that time and chance could write a complex computer program, then it is also absurd to think that time and chance could write the genetic code of life on strands of DNA. Richard Dawkins disagrees: "Evolution produces such a strong illusion of design it has fooled almost every human who ever lived."[17] Statements like this can propel even atheists to refer to Neo-Darwinism as "a heroic triumph of ideological theory over common sense."[18]

Can Evolution Explain Our Origin?

The theory of evolution hypothesizes that all living things came from a simple single cell which originated in a primordial soup. The origin of first life from non-living chemicals (chemical evolution) is a puzzling problem for evolutionary scientists. The 1953 Miller-Urey experiment tried to show that organic molecules could come from inorganic precursors. Simple chemicals were run through a spark chamber, and most of the products they obtained amounted to a brown tar because random chemical reactions produced random molecules. They found a few amino acids, but in a mixture of left-handed and right-handed forms incompatible with life. Only left-handed amino acids are found in the body. Neither nucleotides nor large biomolecules (also essential to life) were produced.[19]

Science has had decades of time and great equipment to produce life in a laboratory but hasn't been able to do so.

Imagine the irony of a frustrated evolutionary scientist after repeated failures to produce life, saying, "If I can just synthesize life here in the lab, then I'll have proven that no intelligence was necessary to form life in the beginning!" Many evolutionists admit that generating life from non-living chemicals is an insurmountable hurdle. So they've resorted to the idea that maybe life came from outside the earth. This is called panspermia and has two versions: 1) Undirected panspermia postulates that chemical evolution happened somewhere, and life was naturally seeded onto the earth somehow. This doesn't actually solve the problem of life from non-living chemicals. It just puts it into the unknown—beyond science. 2) Directed panspermia proposes that intelligent aliens seeded life from outer space onto earth. This tacitly concedes that life was intelligently designed. Conveniently, the designer is an alien who imposes no moral obligations on earthlings. Both concepts are groundless. If evolution by natural selection could not have started in the first place, it's pointless to talk about selection between two runners if both are dead at the starting line![20]

Former atheist Frank Pastore became weary of telling people: "We know God doesn't exist. Therefore, since we're here, it had to have happened this way. Thus, like the universe itself, life, mind, and morality all just popped into existence out of nothingness." Referring to what he called "the four Big Bangs," Pastore said that four questions pertaining to non-evolvable origins that evolution will never be able to answer are: 1) the Cosmological (the universe just popped into existence out of nothingness), 2) the Biological (life just popped into existence out of a dead thing), 3) the Psychological (mind just popped into existence out of a brain), and 4) the Moral (morality just popped into existence out of amorality). After twenty-seven years of atheism, Pastore admitted: "But, for me, the real value of atheism lies in bolstering belief in God. When I doubt, I can begin to doubt my doubts by returning to the Four Big Bangs. And, I eventually fall to my knees and worship, 'In the beginning, God.'"[21]

Is Evolution Compatible With Irreducible Complexity?

Scientists have learned that life on a cellular level requires three essential ingredients: 1) meaningful information, 2) communication channels or networks (including a post office with specialized molecules delivering packs of material to pre-specified destinations within the cell), and 3) a language (for translating between two different worlds: the linear code of the DNA world to the three-dimensional code of the protein world).[22] Information, communication and language are nonmaterial entities that arise through intelligence and are mutually defining (one cannot exist without the others). Such irreducible complexity is the polar opposite of an evolutionary model from a simple cell.

Natural selection can be understood in a developed system. However, it fails to explain how the system developed in the first place. For example, enzymes essential to life are required to read the information coded on DNA. However, the instructions to build the enzymes are on the DNA itself. This chicken-and-egg problem of which came first, enzymes or DNA, illustrates how random processes cannot compete with intelligence. Evolution only makes sense when thinking of it as an assembly of chemicals and structures. However, it loses meaning as an information processing machine.

Molecular biologist Michael Denton explains, "To grasp the reality of life as it has been revealed by molecular biology, we must magnify a cell a thousand million times until it is twenty kilometers in diameter and resembles a giant airship large enough to cover a great city like London or New York. What we would then see would be an object of unparalleled complexity and adaptive design. On the surface of the cell we would see millions of openings, like the port holes of a vast ship, opening and closing to allow a continual stream of materials to flow in and out. If we were to enter one of these openings we would find ourselves in a world of supreme technology and bewildering complexity."

Dr. Denton continues, "Is it really credible that random processes could have constructed a reality, the smallest element of which—a functional protein or gene—is complex beyond our own creative capacities, a reality which is the very antithesis of chance, which excels in every sense anything produced by the intelligence of man? Alongside the level of ingenuity and complexity exhibited by the molecular machinery of life, even our most advanced artifacts appear clumsy... It would be an illusion to think that what we are aware of at present is any more than a fraction of the full extent of biological design. In practically every field of fundamental biological research, ever increasing levels of design and complexity are being revealed at an ever-accelerating rate."[23] To Dr. Denton's point, if random processes could not have created our complex civilization which is the product of human ingenuity, how could random processes have created, developed, and maintained life which is beyond human understanding?

During my career in ophthalmology, I learned that sight could not have evolved from non-sight because of the irreducible complexity of visual system components required to function simultaneously. The eye has 107 million cells and is connected to the brain by over one million neurons. More perfect than any camera ever invented, it caused Charles Darwin to humbly admit, ". . . that the eye with all its inimitable contrivances . . . could have been formed by natural selection seems, I freely confess, absurd in the highest sense."[24]

In attempting to explain the eye's evolution, Richard Dawkins said, "An ancient animal with 5 percent of an eye might indeed have used it for something other than sight, but it seems to me as likely that it used it for 5 percent vision … So, 1 percent is better than blindness. And 6 percent is better than 5, 7 percent better than 6, and so on up the gradual, continuous series."[25] To the contrary, five percent of an eye is not an eye. The same applies to ten, fifty, and seventy-five percent. In fact, the world is full of people with fully formed eyes that cannot perceive light. For any light perception to occur (not to mention focusing), there must be a brain, optic nerve, retina, choroid, and host of other structures that are developed, functioning, and coordinated simultaneously. No credible case can be made for the eye to have evolved randomly.

WHAT IS MACROEVOLUTION?

Macroevolution means evolution across species lines whereas microevolution refers to evolutionary changes within a given species. Unfortunately, the definition of the word "species" is confusing and controversial. Darwin himself said, "…I was much struck how entirely vague and arbitrary is the distinction between species and varieties."[26] Species classification can be based on sexual reproduction, or karyotype (chromosome complement), or DNA sequence, or morphology (form and structure), or behavior, or ecological niche. Paleontologists use the concept of chronospecies (comparative anatomy of fossils) since fossil reproduction cannot be examined.

In the eighteenth century the founder of modern classification, Carl Linnaeus, wanted to name and classify biological species according to Adam's taxonomy of biblical kinds.[27] Each biblical kind (capable of reproducing within itself) would therefore have corresponded to a distinct biological species. A biological species was a population of organisms that could interbreed to produce fertile offspring but could not breed with other biological species. However, over time the meaning of biological species diverged from that of biblical kinds.[28] Today, many different species produce fertile offspring such as false killer whales and dolphins (wholphin), donkeys and zebras (zedonk), polar bears and grizzlies (pizzly), lions and tigers (liger). Darwin's finches with different beaks came from the mainland and could interbreed. They were labeled as thirteen different species even though they diverged from one species.

Classifying new species irrespective of their potential for generating fertile offspring has resulted in over 8 million species on the earth today (85 percent of which have been classified in the 21st century). This figure should continue to escalate since only a small fraction of species on Earth (~14 percent) and in the ocean (~9 percent) have been indexed in a central database.[29] Imagine two people arguing over macroevolution—one mulling over beak sizes while the other ponders the plausibility of non-human ancestors. Today's reference to the term "macroevolution" is neither macro (across reproductive kinds) nor evolution (planned development toward a better system). It simply means an adaptive change in any direction across the line of a recently classified species, be it ever so subtle.

Humans share the highest percentage of identical genomes (genetic content) with chimpanzees. This shouldn't surprise us since humans and chimpanzees are substantially similar anatomically and physiologically. Does this suggest that we have a common ancestor . . . or does it suggest a common architect? Earlier reports that humans and chimpanzees differed genetically by just 1-2% only considered single-letter differences. The inclusion of multi-nucleotide differences ("indels") and whole genomes in the analysis shows the difference to be closer to 10% (about 300 million letter changes).[30] This represents uniquely human-specifying information as is required to program a human to be a human, including a vastly superior brain that is singular among all life—hence, the name *Homo sapiens* ("man" + "thinking"). As different as humans might look and think from each other, if significant mutations like sickle cell anemia, hereditary blindness, etc. were excluded, it would be possible to put all the world's human genomic diversity into a single human couple.[31]

Biblical kinds have diversified widely over time. Red wolves and grey wolves are designated as different species today but they came from a common ancestor. These two can interbreed (along with dogs, coyotes, dingoes and jackals). A kind can interbreed within itself regardless of how its constituents look. Marine iguanas and land iguanas look different, act different, eat different things, and live in different environments. They are labeled as two distinct species by evolutionists who claim they separated millions of years ago. Yet, hybrid offspring of the two species exist, and they are easy to find.[32]

The fossil record shows an abundance of cases which evolutionists describe as "evolutionary stasis" (an oxymoron). These are organisms that have either not changed or changed very little over very long geologic time frames. The "oldest" fossils on earth, stromatolites (estimated to be 3.5 billion years old) are virtually identical to their modern counterparts—blue-green algae. Since stromatolites are found on opposite sides of the world (Australia and the Bahamas), we can confidently infer that they have been subjected to significant environmental changes throughout geologic time—all the more reason for evolution. Yet it did not happen. In Utah, there is a jellyfish well-preserved and unchanged for reportedly half a billion years. While this jellyfish has remained unchanged through the entire evolutionary history of multi-cellular life on earth, according to evolutionists, one of its cousins went on to evolve into humans.[33] Besides claiming to explain radical evolution over time, evolutionists claim to explain "radical stasis" over time.[34]

To express the difficulty paleontologists have of placing bones, tracks, or traces of extinct organisms into discrete evolutionary categories, Professor Maciej Hennenberg wrote: "There is no precise way to test whether Julius Caesar and Princess Diana were members of the same species, *Homo sapiens*."[35] If two humans separated by only 2,000 years cannot be confirmed to be members of the same species, imagine the guesswork involved in creating an evolutionary tree diagram for creatures of different species alleged to have been separated by millions of years! Boundaries between fossil species are extremely mobile, so fossils are moved up and down the evolutionary ladder according to new discoveries and theories.[36]

A number of different versions of evolutionary trees of human origins exist. Conspicuously absent in all of them is a trunk (single common ancestor). Notwithstanding, evolutionists continue to defend the ideology of universal common ancestry. Evolutionary biologist W. Ford Doolittle doubts "there ever was a single universal common ancestor," but "this does not mean that life lacks universal common ancestry because common ancestry does not entail a common ancestor." Instead of conceding what the evidence shows, Dr. Doolittle freely admits that "much is at stake socio-politically," namely the need to defeat "anti-evolutionists" in "the culture wars."[37]

Over the last fifty years, scientific evidence has pointed to an orchard of trees (compatible with biblical kinds) as the origin of life. Members of each kind exhibit rich genetic diversity and the capacity to moderately change in order to adapt and survive in different environments. They are also capable of physiological adaptation through selection as well as variation due to mutations. As long as two modern creatures can hybridize with true fertilization, they are descended from the same kind (the same tree in the orchard). But it does not necessarily follow that if hybridization cannot occur, they are not members of the same kind. Degenerative mutations might be the cause. Couples who cannot have children are not classified as different species or kinds.

While living creatures have been found to adapt to new environments, new kinds have not been discovered. According to marine biologist Robert Carter: "We see a lot of variation potential in nature but real novelties are not there. Different species and speciation in nature is something we can observe, but not novel structures or novel information."[38] Genetic limits have always prevented the formation of new kinds. Professional breeders using intelligent directional artificial selection have not been able to create new kinds. Can we reasonably expect non-intelligent random natural selection to be able to do it?

In the words of Ron Carlson: "In grammar school they taught me that a frog turning into a prince was a fairy tale. In the university they taught me that a frog turning into a prince was a fact!"[39] If you'll continue to study evolution, you'll discover that they were right all along in grammar school.

Is Evolution Compatible With Genetics?

Throughout life we see the effects of mutations all around us. Mutations (changes in our genome) hurt people. They contribute to birth defects, genetic diseases, cancer and aging. Every one of us is a multiple mutant. In our bodies, each of us has about three new mutations every time a cell divides. After continuous cell divisions, a sixty-year-old person has up to 50,000 mutations per skin cell. Total mutations number in the trillions. We transmit a certain fraction of our mutations to our children, and they add more mutations to it and pass it on to the next generation.

Health policies and personal health regimes are aimed at minimizing mutations to reduce the risk of cancer and other degenerative diseases. The devastating effects of mutations make it hard to see any good in them. And yet, the primary axiom of Neo-Darwinism says that mutations are good because they produce the variation and diversity which allow selection and evolution to occur, thus creating the information needed for life.[40]

In recent years I have become interested in the contributions of John C. Sanford. After obtaining his PhD in plant breeding and plant genetics at the University of Wisconsin, Dr. Sanford became a professor at Cornell University for over 30 years during which time he published over 100 scientific presentations and was granted several dozen patents. An evolutionist until the age of fifty, Dr. Sanford trusted science's Neo-Darwinian model even though he didn't understand its mechanics. Initially he thought that "this abstract and highly mathematical field was beyond his own ability to assess critically."[41] Subsequently he discovered that his colleagues and other scientists were also unaware of any scientific basis for the Neo-Darwinian claim.

In time he came to learn that Neo-Darwinism was an axiom . . . a concept that was not testable but accepted by faith because it seemed obviously true to all reasonable parties. He could not find it critiqued in any serious way, either in graduate classes or in graduate level textbooks, or even in the professional literature. His detection of its vulnerability to critical analysis ran counter to the thinking of modern academia which threatened his standing in the academic world.

Still Dr. Sanford persisted with his investigation which led him to the discovery of genetic entropy . . . a degradation of the genome ever since its origin. Experience, genetic theory, numerical simulations, biological data and historical data all pointed him in the same direction toward a biological decay curve of devolution (reverse evolution) for humans along with all other major organisms.

Dr. Sanford's interest in population genetics introduced him to a small specialized subfield of genetics employing theoretical and mathematical models that describe how mutations are passed from one generation to the next and how they affect survivors of a population in each generation. Early population geneticists were all committed Darwinists who described life as "pools of genes"[42] which fit their evolutionary model. However, according to Dr. Sanford, human genes (composed of millions of nucleotides) never exist in pools. They exist in massive integrated assemblies within real people. Each nucleotide (four different molecules constituting the rungs of the DNA ladder) is intimately associated with all the other nucleotides within a given person . . . and they are only selected or rejected as a set of all six billion.

Dr. Sanford illustrates this concept by way of a children's story, "The Princess and the Pea." The royal princess (representing selection) cannot sleep because she feels a pea (representing a mutated nucleotide) beneath her bed even though she is lying on 13 mattresses. Children are entertained by the story because it is so silly. Royalty or not, no one can feel a pea through 13 mattresses. In Dr. Sanford's story adaptation, "The Princess and the Nucleotide Paradox,"[43] the 13 mattresses represent the body's self-regulating homeostasis which prevents pea sized mutations from being felt by the princess's body. Mutations occur at the level of the pea (nucleotide) but selection occurs at the level of the princess (whole organism). Mother Nature (natural selection) never recognizes individual nucleotides. She has to accept or reject a complete set of nucleotides (all or none) which severely limits evolution's potential.

Neo-Darwinism's primary axiom hinges on the genome building capacity of the random mutation/selection model. It has no alternative. A few of the barriers to upward evolution include the following:

- **"Fisher's Fundamental Theorem of Natural Selection"** was pivotal in establishing Neo-Darwinian theory. It portrayed a bell-shaped curve view of mutation, with half of mutations being beneficial and half being deleterious. Sir Ronald Fisher believed that the net effect of mutations would be neutral. However, geneticists report that for every beneficial mutation there are between one thousand and one million deleterious mutations.[44] Dr. Sanford thinks it's closer to one million. Beneficial mutations are so rare that they are typically not even shown in mutation graphs.

- The overwhelming majority of mutations are neither major nor minor but are only "slightly deleterious" (near-neutral) which renders them immune to selection (**Kimura's "no-selection zone"**).[45] Their unstoppable accumulation over time led one prominent population geneticist to ask, "Why have we not died 100 times over?"[46] Some evolutionists have offered the possibility that the effect of many near-neutral mutations occurring together might be great enough for them to be selected out ("synergistic epistasis"). However, interacting mutations (epistasis) create noise that strongly interferes with selection. Numerical simulations using a sophisticated population genetics model called "Mendel's Accountant" shows that synergistic epistasis actually accelerates degeneration and extinction.[47]

- **DNA sequences are poly-functional**, meaning they encrypt multiple overlapping codes (up to 12 each). This makes them poly-constrained. Like a multi-dimensional crossword puzzle, any beneficial mutation in a given location simultaneously disrupts overlapping messages which can only result in degeneration.[48] This type of data compression is overwhelming evidence of design.

- The human genome is composed of physically linked clusters, tying beneficial and deleterious mutations together in linkage blocks. Natural selection cannot separate good and bad mutations. So, the predominantly bad mutations overwhelm the rare good mutations—resulting in a downward ratchet mechanism ("**Muller's Ratchet**"),[49] killing emerging genes long before they can become functional. While selection slows down degeneration, it is powerless to halt it. Dr. Robert Carter illustrates: "Think of a room full of people. Kill off only the ones with the worst or most obvious mutations. What do you have left? A room full of people that still have 60-100 more mutations than their parents. If everyone is a multiple-mutant, and if every generation is more mutant than the one before it, all selection can do is slow down the degeneration by killing off the absolute worst of the lot. But it doesn't stop mutations increasing in the population over time."[50]

- A selectable beneficial mutation must happen about 100 times before it is likely to become amplified and fixed (catch hold) within a population. Geneticist J.B.S. Haldane calculated that it would take 300 generations (6,000+ years) to select a single new mutation to fixation. Selecting one mutation for fixation restricts the ability to select other mutations at the same time (selection interference). This finding, confirmed by others, exposes an unrealistic waiting time for evolution (**"Haldane's dilemma"**).[51] A maximum of 1,000 out of the hundreds of millions of needed beneficial mutations could have possibly become fixed by the time the 6 million year ape-to-human evolution was to have occurred. Moreover, it would have taken over 18 billion years—greatly exceeding the time since the Big Bang—for fixation of the first eight nucleotides (representing an information content of no more than a single word such as "no" or "yes").[52] Evolution's hoped-for ally of time turns out to be its worst enemy. Even in an evolutionary timescale, beneficial fixations could have never created a single gene while deleterious fixations would have mutated humans

> not just to below apes . . . but to extinction, along with apes.[53]

In his own area of plant breeding, Dr. Sanford learned that even directed mutations assisted by selection are unsuccessful in creating beneficial crop changes. Several decades of crop improvement research during the last century using powerful mutagenic agents to create billions of mutation events on millions of plants resulted in a huge number of small, sterile, sick, deformed, aberrant plants. Almost no meaningful crop improvement occurred. The effort was for the most part an enormous failure and was almost entirely abandoned in spite of a host of PhD scientists trying to help it along. However, the same scientists who failed at mutation/selection were extremely successful in crop improvement when they abandoned mutation breeding and instead used the pre-existing natural variation within each plant species or genus.[54]

Dr. Sanford does not doubt there are beneficial mutations that create information, but he is clear that they are exceedingly rare—much too rare for genome building. Instead of producing a net gain of information, beneficial mutations only modulate or fine-tune the body's existing systems. Like switches used to dim and brighten lights, they adjust circuits that are already in place, often doing more damage than good in the process. For example, when a trait called sickle-cell anemia arose in Africa, it allowed people to survive malarial infections. It was a new trait. But the hemoglobin gene was broken in the process. Likewise, antibiotic-resistant bacteria survive because the transporter gene is broken so the poison can't get in. In this artificial environment, bacteria can survive until the antibiotic is stopped. Then the mutant strain is replaced by a superior one. These loss-of-function events represent adaptive degeneration.[55] The predominance of deleterious mutations guarantees the systematic destruction of biological information which is not how genomes are built.

For many decades geneticists have worried about the impact of mutations on the human population. When the concerns first arose, they were based upon an estimated rate of 0.12 to 0.30 deleterious mutations per person per generation. A longstanding belief held that if that figure was found to be as high as 1.0, long-term genetic deterioration would be a certainty. Today the widely accepted rate of mutations per person per generation ranges from 75 to 300.[56] No amount of selection can stop this genomic degeneration together with its loss of genetic information.

Genetic entropy is regularly seen in the aging of our bodies. We can repair teeth, do face lifts, even replace hearts. But it is the cumulative aging of individual cells (principally due to slightly deleterious or near-neutral mutations) that sets an upper lifespan for us. We're mortal. Individually, a cell is trivial and expendable. Collectively, our cells are rusting out like a car and selection cannot stop it. In its final stages, it leads to declining fertility (already down 50% over the past 70 years) which curtails further selection. Eventually, an irreversible downward spiral occurs in a "mutational meltdown" which threatens all advanced species, including mankind.[57]

Genetic entropy has been plotted backwards over a period of 300 generations (6,000+ years). According to evolutionary population geneticist J. F. Crow, the human race is degenerating at a rate of 1-2% per generation due to the accumulation of mutations.[58] The resulting pattern renders a classic biological decay curve, notwithstanding advances in modern medicine, sanitation, and diet. The slope of life expectancy follows a best fit line back to early biblical patriarchs, corroborating biblical data as to their longevity.[59]

Dr. Sanford concludes, "Genetic entropy is profound not only because it impacts us, our children, and our grandchildren. It's really profound because it is lethal, absolutely lethal, to genetic evolutionary theory. It means that things are going down, not up."[60] All modern scientific evidence points to the decay and destruction of an original good design—good information getting worse. But where did this good information come from in the first place? How did life begin? Even under an evolutionary scenario, the first cell would have to have been horrendously complex.

Is Evolution Credible?

Ever since *Genetic Entropy* was first published in 2005, Dr. Sanford has extended numerous invitations for feedback and open dialogue. Appendix 6 of the 2014 edition addresses academia's deafening silence to his requests. Neither substantive rebuttals nor recognitions have been forthcoming. One geneticist friend of Dr. Sanford (a committed evolutionist) confided, "They do not have answers." In fact, many of them had quietly acknowledged in their own papers all of the problems outlined in *Genetic Entropy*. **Mutations do not create us . . . They kill us!** Academia's unwillingness to admit this inescapable conclusion from its own research raises more red flags than a communist party parade as to evolution's credibility. Burying evidence is an effective form of censorship and control. However, the interest of science is never served if its details are treated as trade secrets . . . unavailable for public consideration.

Coordinated attempts to ignore essential evidence is not confined to genetic entropy. Chemist James M. Tour has invited anyone who will to explain the chemical details of macroevolution to him. Dr. Tour makes molecules for a living. He doesn't just buy a kit and mix them. He makes them from scratch which is much harder. He doesn't understand evolution. He has sat in the back rooms of science with National Academy members, with Nobel Prize winners, and with Deans of Departments. Dr. Tour also cannot understand the confidence of scientists who expose their positions on evolution publicly and then become ever so timid when talking with him privately. To them he says, "I understand a lot about making molecules. I don't understand evolution. Do you understand this?" Every time they say, "Uh-uh. Nope." Or, they say nothing. They just stare at him because they can't sincerely respond. So, Dr. Tour posted on his website an interesting offer: "I will buy lunch for anyone who will sit with me and explain evolution, and I won't argue with you until I don't understand something — I will ask you to clarify. But you can't wave by and say, 'This enzyme does that.' You've got to get down to the details of where molecules are built, for me. Nobody has come forward." The Atheist Society contacted him and offered to buy lunch. However, when the Society challenged its members to go to Houston to explain evolution to Dr. Tour, no one volunteered! Dr. Tour concluded, "I sincerely want to know [how macroevolution works.] I would like to believe it. But I just can't."[61]

C.S. Lewis observed: "If the solar system was brought about by an accidental collision, then the appearance of organic life on this planet was also an accident, and the whole evolution of man was an accident too. If so, then all our thought processes are mere accidents—the accidental by-product of the movement of atoms. And this holds for the materialists' and astronomers' as well as for anyone else's. But if their thoughts—i.e., of Materialism and Astronomy—are merely accidental by-products, why should we believe them to be true? I see no reason for believing that one accident should be able to give a correct account of all the other accidents."[62]

Is Evolution Ethical?

People's ethics are an undeniable reflection of their belief system, and how they live their daily lives is profoundly influenced by where they think they came from. Many critics of Christianity point to religious wars, the Crusades, the inquisitions, etc. Those who were acting contrary to the teachings of Christ killed millions of people throughout recorded history.

However, the 20th century suffered far greater genocide at the hands of Hitler, Stalin, Chairman Mao, and Pol Pot, all of whom professed a devout belief in evolution.[63] Mao (who killed 78 million people) regarded himself as a disciple of Darwin. Stalin (who killed 23 million) became a Darwinian convert after reading *On The Origin of Species* at age thirteen. Of Hitler (who killed 17 million), evolutionist Sir Arthur Keith wrote: "The German Fuhrer, as I have consistently maintained, is an evolutionist; he has consciously sought to make the practice of Germany conform to the theory of evolution."[64]

How ironic it is that science is now used as a weapon against the very Christianity which gave it birth. Most branches of science were founded by believers in biblical creation. Indeed, the fathers of modern science were inspired to seek after evidence to bolster their faith. Some atheists admit that science was a child of Christianity but now claim that it is time science grew up and cut the apron strings. However, UK Prime Minister Margaret Thatcher answered that claim: "If you try to take the fruits of Christianity without its roots, the fruits will wither. And they will not come again unless you nurture the roots."[65]

Many evolutionists chide creationists not because of the facts but because creationists refuse to play by the current rules of the game that exclude supernatural creation before examining the evidence.[66] Evolutionary biologist Richard Dickerson declared, "Science is fundamentally a game. It is a game with one overriding and defining rule, Rule #1: Let us see how far and to what extent we can explain the behavior of the physical and material universe in terms of purely physical and material causes, without invoking the supernatural."[67]

One embarrassing result of Rule #1 occurred decades ago when scientists labeled overlapping DNA codes as "Junk DNA" or "selfish genes." Since less than three percent of the genome codes for protein, they concluded that the rest of unimportant leftover garbage was from our evolutionary history. Later the ENCODE project[68] showed that the other ninety-seven percent of the genome was not Junk DNA at all but was very active in creating RNA and controlling many other functions in the cell.[69] In the words of ENCODE director Dr. Ewan Birney: "It is likely that [our current findings of] 80 percent [of functional DNA] will go to 100 percent."[70] At a meeting of the Society for Molecular Biology, evolutionary geneticist Dan Graur said, "If ENCODE is right, then evolution is wrong."[71] Dr. Sanford provides an appropriate elegy for Rule #1: "A lot of scientists now understand that the Junk DNA paradigm was profoundly wrong and will be recorded in history as one of the great blunders in science . . . and it was driven by an ideological commitment to the Darwinian concept."[72]

After World War II the leading Nazis were put on trial. Some claimed they did nothing wrong because the laws of their country said it was okay to kill Jews. So, on what grounds can you put them on trial unless there's a higher morality than national law? Moreover, where can this morality come from if not from the creator of humanity? If we are just rearranged pond scum, there's no such thing as a higher law. You might say that this is stretching things a little, but what if you were one of those consigned to the gulags or the gas chambers or the firing squads, and you also believed in evolution? What basis would you have for saying that they were wrong or acting inconsistently with what both of you believe? Isn't it disingenuous to criticize Christians for breaking their laws but condone evolutionists for doing the same thing in conformity to theirs?

Evolutionists have no objective basis for morality from within their own system. Dr. Dawkins himself admits that "our best impulses have no basis in nature."[73] His fellow non-theistic evolutionary biologist William Provine said that evolution means, "There is no ultimate foundation for ethics, no ultimate meaning in life, and no free will for humans, either."[74] What is our society capable of doing if we fully adopt the evolutionary worldview of deriving our morality from the animal kingdom? Such a thought should cause a revulsion in us! Over the last 100 years, almost all of biology has proceeded independent of evolution, except evolutionary biology itself.[75]

The National Academy of Sciences (NAS) has published an educator's guidebook entitled, "Teaching about Evolution and the Nature of Science." Distributed free of charge to public schools and educators, it persuasively and professionally presents its goal of extinguishing belief in creation. Many people do not realize that evolution propagates an anti-biblical religion. The first two tenets of the "Humanist Manifesto I," signed by many prominent evolutionists, are: "1) Religious humanists regard the universe as self-existing and not created. 2) Humanism believes that Man is a part of nature and has emerged as a result of a continuous process."[76]

Many humanists are quite open about using the public schools to proselytize their faith. This might surprise some parents who think schools are free of religious indoctrination. The following statement corrects such a misunderstanding:

"I am convinced that the battle for humankind's future must be waged and won in the public school classroom by teachers who correctly perceive their role as the proselytizers of a new faith: a religion of humanity that recognizes and respects the spark of what theologians call divinity in every human being. These teachers must embody the same selfless dedication as the most rabid fundamentalist preachers, for they will be ministers of another sort, utilizing a classroom instead of a pulpit to convey humanist values in whatever subject they teach, regardless of the educational level—preschool, daycare, or large state university. The classroom must and will become an arena of conflict between the old and new—the rotting corpse of Christianity, together with all its adjacent evils and misery, and the new faith of humanism ... It will undoubtedly be a long, arduous, painful struggle replete with much sorrow and many tears, but humanism will emerge triumphant. It must if the family of humankind is to survive."[77]

If these tenets are successfully incorporated into classrooms, what resources will we have to help our children when they become depressed, suicidal, or violent? Evolution is promoted as unpredictable, natural, and without a specific direction or goal. However, the ideology that sponsors it is predictable, unnatural, and with a specific direction and goal.

1. Paul Johnson, attorney and student of science said: "There is ample reason to believe that Darwinism is sustained not by an impartial interpretation of the evidence, but by dogmatic adherence to a philosophy even in the teeth of the evidence." Interestingly, there is also a real danger in the world that if you stray from the company line about Darwinism, you are in danger of losing your job.[78]

2. J.Y. Chen, Chinese paleontologist, issued the following statement during one of his lectures in the United States: "In China, we can criticize Darwin,

but not the government. In America, you can criticize the government, but not Darwin."[79]

3. Colin Patterson, senior paleontologist of the British Museum of Natural History, in his book *Evolution Explained*, was questioned as to why he did not include any pictures or illustrations of transitional forms. He said: "If I knew of any, fossil or living, I would certainly have included them… I will lay it on the line—there is not one such fossil for which one could make a watertight argument."[80]

4. Niles Eldredge of the American Museum of Natural History was even bolder when he admitted: "We paleontologists have said that the history of life supports [the story of gradual adaptive change] knowing all the while it does not."[81]

5. David Berlinski said: "The greater part of the debate over Darwin's theory is not in service to the facts. Nor to the theory. The facts are what they have always been: They are unforthcoming. Among evolutionary biologists, these matters are well known. In the privacy of the Susan B. Anthony faculty lounge, they often tell one another with relief that it is a very good thing the public has no idea what the research literature really suggests.

'Darwin?', a Nobel laureate in biology once remarked to me over his bifocals. 'That's just the party line.'"[82]

6. George Wald, professor emeritus of biology at the University of Harvard, might have summed it up best: "There are only two possibilities as to how life arose. One is spontaneous generation arising to evolution; the other is a supernatural creative act of God. There is no third possibility. Spontaneous generation, that life arose from non-living matter was scientifically disproved 120 years ago by Louis Pasteur and others. That leaves us with the only possible conclusion that life arose as a supernatural creative act of God. I will not accept that philosophically because I do not want to believe in God. Therefore, I choose to believe in that which I know is scientifically impossible: spontaneous generation arising to evolution."[83]

As of April 2020, more than 1,100 scientists and researchers in chemistry, biology, medicine, physics, geology, anthropology, paleontology, statistics, and other fields signed a statement of scientific dissent from Darwinism. It read: "We are skeptical of claims for the ability of random mutation and natural selection to account for the complexity of life. Careful examination of the evidence for Darwinian theory should be encouraged."[84] However, if you do an internet search for "evolution," you won't find this dissent or any other challenges to the evolution theory in at least the first twelve pages. If you are a student, evolution will be taught to you as a proven theory. If you are a researcher, you know that publishing a study challenging evolution could threaten your career. On Wikipedia, any theories about the origin of life that differ from evolution are labeled as "pseudoscience." But is evolution, in fact, the greatest show on earth[85] as Richard Dawkins claims, or is it the greatest hoax on earth[86] as Jonathan Sarfati counterclaims?

Is Evolution Science Or Myth?

Evolution's ideological implications are so powerful that people's interpretation of evolutionary evidence often reveals more about themselves than about the evidence. If science were personified, it wouldn't care about the source of its evidence . . . only about the weight. Surely science laughs in derision at its guardians who categorically dismiss evidence from detestable sources.

The case for evolution is based on biological hardware (physical systems) since biological software is so irreducibly complex as to defy explanation for an evolvable origin. By assuming that similarities in organisms suggest random evolutionary changes over time, evolutionists overlook the fact that creative artists (including divine ones) have their own unique styles. However, evolution's silver bullet is genetic entropy from which it has no defense. The reality is that humans along with all other complex living organisms are devolving, not evolving. Since evolution and devolution cannot co-exist, science dictates that proof of devolution makes its opposite, evolution, a myth.

However, this particular myth is too psychotropic to rest in peace. The elixir of evolution is more resistant to prohibition than was alcohol a century ago. In a world of limitless information, biological devolution is being craftily concealed while its reporters are vilified as creationists. Ardent evolutionists continue to show broken organisms as examples of adaptive mutations and natural selection to try to convince us that speciation really is evolution. Complicit with evolution's ideology is its sedating secularism of a no-host world . . . a powerful induction to the anesthesia of atheism which desensitizes the conscience ("with" + "science") from distinguishing between good and evil. This is not the science I learned and loved as a child!

In contrast to evolution's godless creation, purposeless existence, and absent future, devolution opens the door to the realization that our lives are time-sensitive gifts linked to opportunity, growth, and destiny . . . but from whom? If life were the result of a seeded primordial soup, how could we ever thank our alien benefactors? Wouldn't it be reasonable for them to be interested in our welfare?

Logic suggests that they might have wanted to form us in their image, nurture us, cultivate our love, communicate their reason for creating us and give us a vision of our destiny. In truth, evolutionism is as mythical as religion as it is as science.

What Is Devolution?

Carl Sagan popularized the dictum, "Extraordinary claims require extraordinary evidence." He was actually rewording Laplace's principle which says that "the weight of evidence for an extraordinary claim must be proportioned to its strangeness."[87] Devolution is an extraordinary claim but the evidence for genetic entropy is proportionately extraordinary and strange! Even more compelling than Copernicus' 1543 discovery that the sun is the center of the universe, devolution is a powerful reminder to us that the Son is the center of the universe . . . the Son of God!

In order to understand devolution, the "devol" is in the details. Synonyms for devolution include deterioration, degeneration, descent, decline, retrogression . . . and fall . . . which turns our attention to the gospel of Jesus Christ. Holy writ tells us: "For as in Adam all die, even so in

Christ shall all be made alive."[88] The Fall of Mankind is a divine explanation for the scientific observation that we are devolving. . . but for what purpose?

The Apostle Peter taught, "Beloved, think it not strange concerning the fiery trial which is to try you, as though some strange thing happened unto you: But rejoice, inasmuch as ye are partakers of Christ's sufferings; that, when his glory shall be revealed, ye may be glad also with exceeding joy."[89] Our trials are part of our individualized curriculum for growth in mortality. They also direct us to our Redeemer, Jesus Christ, whose incomprehensible suffering on our behalf qualifies Him to commiserate with our deepest anguish.

Peter goes on to say: ". . . though now for a season, if need be, ye are in heaviness through manifold temptations: That the trial of your faith, being much more precious than of gold that perisheth, though it be tried with fire, might be found unto praise and honour and glory at the appearing of Jesus Christ: Whom having not seen, ye love; in whom, though now ye see him not, yet believing, ye rejoice with joy unspeakable and full of glory: Receiving the end of your faith, even the salvation of your souls."[90] Trials of faith are our best preparation for eternity. Although results are mixed, success invariably follows those who persevere in faith.[91]

In His parable of the sower, Jesus described seeds that either fell 1) by the way side where fowls devoured them up, 2) on stony places where the sun scorched them, 3) among thorns that choked them, or 4) into good ground bringing forth fruit with widely varying degrees of success. Later He explained: "When any one heareth the word of the kingdom, and understandeth it not, then cometh the wicked one, and catcheth away that which was sown in his heart. This is he which received seed by the way side. But he that received the seed into stony places, the same is he that heareth the word, and anon with joy receiveth it; Yet hath he not root in himself, but dureth for a while: for when tribulation or persecution ariseth because of the word, by and by he is offended. He also that received seed among the thorns is he that heareth the word; and the care of this world, and the deceitfulness of riches, choke the word, and he becometh unfruitful. But he that received seed into the good ground is he that heareth the word, and understandeth it; which also beareth fruit, and bringeth forth, some an hundredfold, some sixty, some thirty."[92]

The parable of the sower references the soil of our heart which is the environment for the seed of our faith. Devolution is life's personalized pruning that purges (selects out) our obstacles to spiritual evolution.

What Is Spiritual Evolution?

Darwin's trunkless evolutionary tree prompts our search for life's true and living *trunk*. Using plant terminology, Jesus meekly declared, "I am the true vine, and my Father is the husbandman. Every branch in me that beareth not fruit he taketh away: and every branch that beareth fruit, he purgeth it, that it may bring forth more fruit. Now ye are clean through the word which I have spoken unto you. Abide in me, and I in you. As the branch cannot bear fruit of itself, except it abide in the vine; no more can ye, except ye abide in me. I am the vine, ye are the branches: He that abideth in me, and I in him, the same bringeth forth much fruit: for without me ye can do nothing. If a man abide not in me, he is cast forth as a branch, and is withered; and men gather them, and cast them into the fire, and they are burned."[93]

I have heard people say (as if to shrug their shoulders), "I'm just not spiritual." What they are really saying is that they are not spiritually healthy. Refusing to acknowledge one's own spirituality is like never exercising and then saying, "I'm just not physical!" Our mortal body isn't us. . . It's ours. Our spirit body (along with our personality) is us—the life of our temporary mortal tabernacle. That is why spiritual fitness is so much more valuable than physical fitness. Our spirit continues to live after our mortal body dies. The extent to which we grow (evolve) spiritually is what determines the quality of our ultimate resurrection (reuniting spirit with body).[94]

To help us understand our spiritual nature, the Apostle Paul said: ". . . we have had fathers of our flesh which corrected us, and we gave them reverence: shall we not much rather be in subjection unto the Father of spirits, and live?"[95] We are not aliens nor products of aliens but literal spirit sons and daughters of God who is our Heavenly Father. The significance of this relationship is profound! Children naturally want to grow up to be like parents who are worthy of emulation. Paul goes on to say: "The Spirit itself beareth witness with our spirit, that we are the children of God: And if children, then heirs; heirs of God, and joint-heirs with Christ; if so be that we suffer with him, that we may be also glorified together."[96] Who wouldn't like to be an heir to a wealthy father? In order to do so, we need to accept the invitation to join our "Father's business . . . to bring to pass the immortality and eternal life of man"[97] (our brothers and sisters).

Born with a purpose and destiny as heirs of God, we need to seek to become like Him by emulating the example of His only begotten Son in the flesh. When we do, the mediocrity of our lives evolves into godliness and our spiritual receptivity increases. Of course, we cannot expect to be perfect (literally complete or finished) of our own volition. The Lord makes up for our deficits through His atoning sacrifice as we commit ourselves to His covenant path. Along the path, as we strive for perfection, we achieve excellence.

The Psalmist said: "Ye are gods; and all of you are children of the most High."[98] This scripture references our spiritual embryological state and confirms our promise of heirship. From a scientific perspective, since we are of the same species as God, our godly evolution is microevolutionary even though everything about it is macro! The expression "heirs of God, and joint-heirs with Christ" does not mean rivalry for God's supreme power and authority. It signifies our qualifying for His society and earning the right to enjoy the blessings of eternity with our families. The Lord has made it clear that there is a place in His kingdom for all who are willing to follow His covenant path.[99]

Our destiny is ultimately determined by our personal choices. This is why the Lord said, "Care not for the body, neither the life of the body; but care for the soul, and for the life of the soul."[100] In other words, don't invest in depreciating assets. Invest in that which contributes to your spiritual growth. Death is a comma, not an exclamation point! To believe that death is the end of one's identity and progress is to miss the universe's astonishing efficiency, particularly with respect to its most valuable asset: mankind, God's greatest creation. Like an acorn that is destined to become a giant oak tree, each of us is a spirit child of celestial parentage with celestial potential. The course is laid before us along with all of the resources we'll ever need through the gospel of Jesus Christ.

How Can We Expedite Our Spiritual Evolution?

Jesus taught, "And this is life eternal, that they might know thee the only true God, and Jesus Christ, whom thou hast sent."[101] Sadly, "more mortals die in ignorance of God's true character than die in actual defiance of Him."[102] Too many people resemble Galileo's contemporaries who were afraid to understand what their eyes saw through his telescope. Our fears would be better served by being aware of those who are unimposing about divine ways but are very serious about imposing their own ways! God is more committed to our freedom than they are! This principle is beautifully expressed in the following hymn:

> "Know this, that ev'ry soul is free
> To choose his life and what he'll be;
> For this eternal truth is giv'n:
> That God will force no man to heav'n.

He'll call, persuade, direct aright,
And bless with wisdom, love, and light,
In nameless ways be good and kind,
But never force the human mind."[103]

Essential to our spiritual evolution is knowing God. Many people are no more aware of Him than are goldfish toward their caretaker. However, God isn't hard to find when we make an effort to reach out to Him. More than any other invitation, the Lord has generously invited us to "Ask, and it shall be given you; seek, and ye shall find; knock, and it shall be opened unto you: For every one that asketh receiveth; and he that seeketh findeth; and to him that knocketh it shall be opened."[104] This should be our top priority in life! Jesus further counsels us to not waste our efforts on too many concerns that will be forgotten tomorrow, ". . . for your heavenly Father knoweth that ye have need of all these things. But seek ye first the kingdom of God, and his righteousness; and all these things shall be added unto you."[105] Indeed they are!

In science we prove, but in faith we discover. Spiritual truths are not discovered by mankind's scientific method. The Prophet Isaiah said, "For my thoughts are not your thoughts, neither are your ways my ways, saith the Lord. For as the heavens are higher than the earth, so are my ways higher than your ways, and my thoughts than your thoughts."[106] Inasmuch as God's methods are higher than ours, they must be studied, learned and applied according to His terms if spiritual evolution is to occur.

Prayerfully asking, seeking and knocking activates our spirit to the Lord's divine channel. Answers to our prayers do not typically come immediately (in line with our schedule) but according to God's timetable. When we are ready, the answers come! Analogous to the three essential ingredients of cellular life are three essential ingredients of spiritual life: 1) meaningful information (from God), 2) a communication channel or network (paired to our spirit), and 3) a language (for translating between two different worlds for our understanding through the Holy Ghost).

Like the cosmos which is over 95 percent dark matter and dark energy, the overwhelming majority of all there is to discover is darkness to science. However, we live in a witnessing world in which wisdom is accessible to us through divine revelation. As we come to understand that one day "every knee shall bow . . . and every tongue shall confess" that Jesus is the Christ, we are moved to ask ourselves, "Why don't I do it now?" Since "every one of us shall give an account of himself to God,"[107] it will mean much less to kneel down when it is no longer possible to stand up. Spiritual evolution is the result of foresight and initiative . . . not of seeing truth too late.

Like parents who speak to their little children in a language they can understand, God communicates to us according to our level of spiritual maturity. Besides personal revelation, He also reaches out to us by means of His prophets. The Old Testament says: "Surely the Lord God will do nothing, but he revealeth his secret unto his servants the prophets."[108] As a result of incredible sacrifices, the words of prophets of old have been preserved for us in Holy Scriptures which should be welcomed and studied regularly while recognizing imperfections in language and transcribing over the millennia. The Lord still speaks today through His living prophets.[109]

The devolutionary effects of the Fall have been overcome by the resurrection and atoning sacrifice of the Lord Jesus Christ. As we nurture our faith, Peter promises that we will "be partakers of the divine nature, having escaped the corruption that is in the world . . . And beside this, giving all diligence, add to your faith virtue; and to virtue knowledge; And to knowledge temperance; and to temperance patience; and to patience godliness; And to godliness brotherly kindness; and to brotherly kindness charity. For . . . if ye do these things, ye shall never fall: For so an entrance shall be ministered unto you abundantly into the everlasting kingdom of our Lord and Saviour Jesus Christ."[110]

CONCLUSION

Austin Farrer said: "Though argument does not create conviction, lack of it destroys belief. What seems to be proved may not be embraced; but what no one shows the ability to defend is quickly abandoned. Rational argument does not create belief, but it maintains a climate in which belief may flourish."[111] My personal search for understanding evolution introduced me to a climate that is not only compatible with, but powerfully conducive to faith. Although I had not intended to demonize biological evolution in the process, the more I studied . . . the more I became convinced of its misrepresentation of science and of the toxic fallout it heaps upon the lives of those who embrace it. Two particular consequences of evolution include its theft of personal gratitude and genuine self-worth.

Gratitude: After realizing the impossibility of explaining evolution's origins, some evolutionists have claimed that the origin of life from non-living chemicals (abiogenesis) has nothing to do with evolution. Their evolutionist colleague Gordy Slack rebuked them for that conclusion: "I think it is disingenuous to argue that the origin of life is irrelevant to evolution. It is no less relevant than the Big Bang is to physics or cosmology. Evolution should be able to explain, in theory at least, all the way back to the very first organism that could replicate itself through biological or chemical processes. And to understand that organism fully, we would simply have to know what came before it. And right now we are nowhere close."[112]

Imagine a little boy who receives from his parents a bicycle with training wheels. The parents help their son onto the seat, teach him how to pedal, and stand by to watch. Within a few feet, they are stunned to hear their son say, "This is a great gadget! It was generated spontaneously, naturally adapts to tilt, and randomly selects from walking." Although this analogy might sound particularly absurd to an evolutionist, it begs the question: "How much does our belief in evolution cause our loving Heavenly Father to weep[113] over us?" It was He who gave us muscles to smile or sneer, as well as breath to praise or profane Him.

Jesus taught the importance of gratitude through His encounter with ten lepers who asked Him to have mercy on them. After healing them all, nine failed to offer any expression of appreciation. However, one of them, a hated Samaritan, glorified God, fell down on his face at Jesus' feet, and thanked Him. This merited the Samaritan a greater blessing than the healing of his leprosy. The Lord said to him: "Arise, go thy way: thy faith hath made thee whole."[114] Jesus was referring to the Samaritan's spiritual wholeness. Contrast this with the plight of an evolutionist who might feel grateful but has no one to thank.

Self-worth: Born within us is a desire to explore our personal roots. Evolution's proposal that our lives are the result of alien visitors or inorganic chemicals is, in fact, an endorsement for us to become self-imposed aliens. To the Ephesians, Paul said: ". . . at that time ye were without Christ, being aliens from the commonwealth of Israel, and strangers from the covenants of promise, having no hope, and without God in the world."[115] A belief in evolution imprisons us well below our privileges. What good is it to be children of God if we ignore our divinely inherited attributes and behave like orphans?

Tragically, for the many who do not choose the narrow way to spiritual evolution, their unconscious selection becomes the broad way to "spiritual evolutionary stasis."[116] Paul described some of the character traits of those who traipse this tumbleweed trail: "This know also, that in the last days perilous times shall come. For men shall be lovers of their own selves, covetous, boasters, proud, blasphemers, disobedient to parents, unthankful, unholy, Without natural affection, trucebreakers, false accusers, incontinent, fierce, despisers of those that are good, Traitors, heady, highminded, lovers of pleasures more than lovers of God; Having a form of godliness, but denying the power thereof . . . Ever learning, and never able to come to the knowledge of the truth."[117]

Elbert Hubbard once said, "I am looking out through the library window into the apple orchard, and I see millions of blossoms that will never materialize or become fruit for lack of vitalization."[118] The proper destiny of an apple blossom is to become an apple. But before that can happen, a bee or some other instrument of vitalization must plant the fertilizing pollen in the blossom's heart to give it its opportunity to reach its destiny. Similarly, the rightful destiny of a human soul is to become even as God. Like the bee, we need to get honey out of the same flowers from which the spider can extract only poison.

We do not need to know everything about evolution in order to turn our understanding of it from poison to honey. We only need to know that it is a spiritual phenomenon. When answers to the deep questions of life do not seem to be forthcoming, our spiritual evolution will nevertheless proceed as long as we follow God's instructions for unlocking the combination to heaven's vault: 1) one tumbler falls when there is sincere faith, 2) a second one falls when there is an effort to keep His commandments, and 3) the third and final tumbler falls when what is sought is according to His will and timing. Pounding on the vault is pointless. Personal revelation requires a security clearance of our real intent.[119] We won't understand everything yet, but we'll know that our Heavenly Father is in control,[120] that He loves us,[121] and that we can trust Him.[122]

Recently, a television broadcaster expressed regret over selling and buying back a certain stock seventeen times from $4 per share in 2010 all the way up to $775 per share in 2021. Citing volatility as the cause of his inability to hold on to the stock, he illustrated a challenge which many of us have in holding on to our faith. Jesus emphasized our need for unwavering faith in His parable of ten virgins in which five wisely kept their lamps full of oil (symbolizing faith) in preparation to meet the bridegroom at his unscheduled coming. The five virgins who were foolish thought they could obtain oil (faith) at the last minute in time for the marriage.[123] However, faith can no more be summoned in a crisis than piano skills can be developed in a concert. Overcoming false notions about evolution is necessary so "that we henceforth be no more children, tossed to and fro, and carried about with every wind of doctrine, by the sleight of men, and cunning craftiness, whereby they lie in wait to deceive; But speaking the truth in love, may grow up into him in all things, which is the head, even Christ."[124]

If you have sold your faith, the best investment advice I can give you is to buy it back, regardless of its cost to you, through authorized personal covenants with the Lord while your mortal probation market is still open. If you have lost your faith, search diligently until you find it again. If you have never had faith, grant yourself permission to be "taken out" of Neo-Darwinian thinking so you won't be afraid of being "taken in" by Christ. Your love for Him will become sweeter than honey! Then hold on to your faith tenaciously in the midst of storms, constantly building upon it through your efforts. As you replace the stony and thorny soil of your heart with enriched soil, there will be "joy…in heaven over [you]."[125] Moreover, you will be sustained in crises through the corridors of your devolutionary mortality until the day in which you learn that "eye hath not seen, nor ear heard, neither have entered into the heart of man, the things which God hath prepared for them that love him."[126]

In the prologue of his book, *Genetic Entropy*, Dr. Sanford writes: "I realize I have wasted much of my life arguing about things that don't really matter. It is my sincere hope that this book can actually address something that really does matter. The issues of who we are, where we come from, and where we are going seem to me of enormous importance. This is the real subject of this book."[127] Dr. Sanford's stated purpose inspires us to ask ourselves: "Is there is any geopolitical, climatological, socioeconomic, or other proximate concern of greater importance than knowing that I am on a divinely-guided path, beckoning me upward to my greatest eternal joy and fulfillment?" Such a question is worthy of our deepest, ongoing contemplation. It is our attitude, not our environment, that directs us toward our ultimate destiny.

George Bernard Shaw could have very well been describing our evolution-indoctrinated world when he said: "In an ugly and unhappy world, the richest man can purchase nothing but ugliness and unhappiness." Remarkably, in that same world even the poorest souls who love God and love His children can afford the finest beauty and greatest happiness. The difference lies in the character we create for ourselves as C.S. Lewis illustrates: "It is a serious thing to live in a society of possible gods and goddesses, to remember that the dullest and most uninteresting person you can talk to may one day be a creature which . . . you would be strongly tempted to worship, or else a horror and a corruption such as you now meet, if at all, only in a nightmare. All day long we are, in some degree, helping each other to one or other of these destinations. It is in the light of these overwhelming possibilities, it is with the awe and circumspection proper to them, that we should conduct all of our dealings with one another, all friendships, all loves, all play, all politics. There are no ordinary people. You have never talked to a mere mortal. Nations, cultures, arts, civilizations—these are mortal, and their life is to ours as the life of a gnat. But it is immortals whom we joke with, work with, marry, snub, and exploit—immortal horrors or everlasting splendours."[128]

Truly, as our lives approach everlasting splendor our world evolves heavenward, as well! In the words of Elder Neal A. Maxwell: "Those who say life is meaningless will yet applaud the atonement which saved us from meaninglessness. Christ's victory over death routs the rationale that there is a general and irreversible human predicament; there are only personal predicaments, but even from these we can be rescued by following the pathway of Him who rescued us from general extinction."[129]

NOTES

1. I attended California Military Academy, 5300 Angeles Vista Boulevard, Los Angeles, California (founded 1906) from 1956-1961 (fifth through ninth grades).
2. Fischer, Martin. *Gracian's Manual: A Truth-Telling Manual and the Art of Worldly Wisdom*. Charles C. Thomas Publishers, 1949, pp. 92.
3. Darwin, Charles. *On the Origin Of Species*. John Murray, 1859.
4. The Scopes Trial, also known as the Scopes Monkey Trial, was the 1925 prosecution of science teacher John Scopes for teaching evolution in a Tennessee public school.
5. Tallentyre, S.G. "Helvétius: The Contradiction." *The Friends of Voltaire*. London: Smith, Elder, & Co., 1906, pp. 199.
6. John 7:17
7. Sanford, John C. *Genetic Entropy*. FMS Publications, 2014, pp. vi-vii.

8. Dawkins, Richard. *The Blind Watchmaker: Why the Evidence of Evolution Reveals a Universe without Design*. New York: W.W. Norton, 1986, pp. 6.
9. Lewontin, Richard C. "Billions and Billions of Demons." *New York Review of Books*. 9 January 1997.
10. Zimmer, C. "The Neurobiology of the Self." *Scientific American*, November, 2005.
11. Marshall, Perry. *Evolution 2.0: Breaking the Deadlock Between Darwin and Design*. Benbella Books, 2015, pp. 36-7.
12. Behe, Michael J. *Darwin Devolves: The New Science About DNA That Challenges Evolution*. Harper One, 2019, pp. 19.
13. Bowler, Peter J. "The Changing Meaning of 'Evolution.'" *Journal of History of Ideas*, University of Pennsylvania Press. Vol 36, No. 1 (Jan.- Mar. 1975) pp. 106.
14. Steels, Luc. "Do languages evolve or merely change?" *Journal of Neurolinguistics*. Published by Elsevier Ltd. 15 November 2016.
15. Geisler, Norman L. and Frank Turek. "New Life Forms: From the Goo to You via the Zoo?" *I Don't Have Enough Faith to Be an Atheist*. Crossway, 2004, pp. 137-8.
16. Johnson, Jeffrey D. "The Irrationality of Evolution." *The Absurdity of Unbelief: A Worldview Apologetic of the Christian Faith*. Free Grace Press, 2016, pp. 145.
17. Dawkins, Richard. *The Rise of Atheism*. 2010.
18. Nagel, Thomas. *Mind and Cosmos: Why the Materialist Conception of Nature Is Almost Certainly False*. New York: Oxford University Press, 2012.

19. Sarfati, Jonathan, "The Origin of Life," *Evolution's Achilles' Heels*. Creation Ministries International, 2014. Film.
20. Sarfati, Jonathan. "Origin of Life." *The Greatest Hoax on Earth? Refuting Dawkins on Evolution*. Creation Book Publishers, 2014, pp. 223.
21. Pastore, Frank. "Why Atheism Fails: The Four Big Bangs." 6 May 2007.
22. Sanford, John C. and Robert Carter. "Genetics," *Evolution's Achilles' Heels*. Creation Ministries International, 2014. Film.
23. Denton, Michael. *Evolution: A Theory in Crisis*. Adler and Adler Publishers, Inc., 1986, pp. 328, 342.
24. Thompson, Bert and Wayne Jackson. *The Case for the Existence of God*. Apologetics Press, 1966.
25. Johnson, Phillip E. *Darwin on Trial*. Regnery Gateway, 1991, pp. 34.
26. Darwin, Charles. *On the Origin Of Species*. John Murray, 1859, pp. 48.
27. Genesis 1:11-12, 21, 24-25; 2:19-20
28. Sarfati, Jonathan. "Species and Kinds." *The Greatest Hoax on Earth? Refuting Dawkins on Evolution*, Creation Book Publishers, 2014, pp. 30-1.
29. Mora, Camilo; Tittensor, Derek P.; Adl, Sina; Simpson, Alastair G. B.; Worm, Boris. "How Many Species Are There on Earth and in the Ocean?" *PLOS Biology*, 23 August 2011.
30. Rupe, Christopher and Dr. John Sanford. *Contested Bones*, FMS Publications, 2019, pp. 311

31. Carter, Robert. "Genetics." *Evolution's Achilles' Heels*. Creation Ministries International, 2014. Film.
32. Batten, Donald. "Natural Selection." *Evolution's Achilles' Heels*. Creation Ministries International, 2014. Film.
33. Salisbury, David. "Could comb jellies, close cousins of the jellyfish, be the earliest ancestors of animals?" *Research News*, Vanderbilt University, December 13, 2013. Andreas Baexevanis, Cornelius Vanderbilt Chair in Biological Sciences, is quoted as saying, "ctenophores are the oldest relative of the entire animal family, including humans."
34. Silvestru, Emil. "The Fossil Record." *Evolution's Achilles' Heels*. Creation Book Publishers, 2015, pp. 139-42.
35. Hennenberg, M., *The Hobbit Trap*, Wakefield Press, Kent Town, Australia, 2008, pp. 25.
36. Silvestru, Emil. "The Fossil Record." *Evolution's Achilles' Heels*. Creation Book Publishers, 2015, pp. 131.
37. Doolittle, W. Ford. "The practice of classification and the theory of evolution, and what the demise of Charles Darwin's tree of life hypothesis means for both of them." *Philosophical Transactions of the Royal Society of London* B 364, 2009: 2221-2228.
38. Carter, Robert. "Natural Selection." *Evolution's Achilles' Heels*. Creation Ministries International, 2014. Film.
39. Geisler, Norman L. and Frank Turek, "New Life Forms: From the Goo to You via the Zoo?" *I Don't Have Enough Faith to Be an Atheist*, Crossway, 2004, pp. 137.

40. Sanford, J.C. *Genetic Entropy*, FMS Publications, 2014, pp. 15-16.
41. Ibid, pp. ii.
42. Ibid, pp. 60.
43. Ibid, pp. 53-4, 72-3.
44. Ibid, pp. 22-24, 32, 182-3.
45. Ibid, pp. 34, 247.
46. Kondrashov, A.S. 1995. *Contamination of the genome by very slightly deleterious mutations: why have we not died 100 times over?* J. Theor. Biol. 175:583-594.
47. Sanford, J.C. *Genetic Entropy*, FMS Publications, 2014, pp. 169-73.
48. Ibid, pp. 140-42.
49. Ibid, pp. 92, 146-7.
50. Carter, Robert. "Genetics." *Evolution's Achilles' Heels*. Creation Ministries International, 2014. Film.
51. Sanford, J.C. *Genetic Entropy*, FMS Publications, 2014, pp. 134-8, 175-6, 246.
52. Rupe, Christopher and Dr. John Sanford. *Contested Bones*, FMS Publications, 2019, pp. 314
53. Sanford, J.C. *Genetic Entropy*, FMS Publications, 2014, pp. 148-9.
54. Ibid, pp. 26-7.
55. Ibid, pp. 28-9.
56. Ibid, pp. 37-8.
57. Ibid, pp. 44-5.
58. Ibid, pp. 71.

59. Ibid, pp. 158-60, 168.
60. Sanford, John C. "Genetics," *Evolution's Achilles' Heels*. Creation Ministries International, 2014. Film.
61. Tour, James M. "Nanotech and Jesus Christ." YouTube, uploaded by the Veritas Foundation, June 2014.
62. Lewis, C.S. "God in the Dock." *Essays on Theology and Ethics*. William B. Eerdmans Publishing Co., 1972, pp. 52-3.
63. Catchpoole, David and Mark Harwood. "Ethics and Morality." *Evolution's Achilles' Heels*, Creation Book Publishers, 2015, pp. 247-8.
64. Keith, Sir Arthur. *Evolution and Ethics*. New York: Putnam, 1947, pp. 230.
65. Thatcher, Margaret. "Christianity and Wealth." Church of Scotland General Assembly. 21 May 1988. Address.
66. Wieland, Carl. "Science: the rules of the game." *Creation*, December 1988.
67. Dickerson, R.E. *J. Molecular Evolution*. 1992; *Perspectives on Science and the Christian Faith*. 1992.
68. Sanford, J.C. *Genetic Entropy*, FMS Publications, 2014, pp. 211. The ENCODE project took over 10 years and cost roughly 400 million dollars. It involved 442 scientists from all over the world and resulted in numerous publications in various journals.
69. Carter, Robert. "Natural Selection." *Evolution's Achilles' Heels*. Creation Ministries International, 2014. Film lists the following non-protein functions of Junk DNA: pervasive alternate splicing, pseudogene RNA interference,

intronic elements, chromatin organization, cell differentiation, L1 gene silencing, X chromosome inactivation, alternate transcription start sites, regulation of transcription speed, pervasive transcription, sequence-independent structures, chromosome looping, and many more.

70. Sanford, J.C. *Genetic Entropy*, FMS Publications, 2014, pp. 21, 212.
71. Rupe, Christopher and Dr. John Sanford. *Contested Bones*, FMS Publications, 2019, pp. 307.
72. Sanford, John C. "Genetics," *Evolution's Achilles' Heels*. Creation Ministries International, 2014. Film.
73. Dawkins, Richard. *The God Delusion*. London: Transworld Publishers, 2006, pp. 97.
74. Provine, William B. Origins Research. 1994, pp. 9.
75. Kirschner, Marc. *Boston Globe*. 23 October 2005.
76. Sarfati, Jonathan. *Refuting Evolution*. Creation Book Publishers, 2019, pp. 20.
77. Dunphy, J. "A Religion for a New Age." *The Humanist*. January/February 1983, pp. 23, 26.
78. Evans, Duncan. "Why Darwinian Theory is Fading." *The Epoch Times*, July 19, 2021.
79. Ibid.
80. Sarfati, Jonathan. *Refuting Evolution*. Creation Book Publishers, 2019, pp. 48.
81. Lennox, John C. *God's Undertaker: Has Science Buried God?* Kregel Publications, 2009, pp. 113-4. Paleontologists around the world have discovered a sudden explosion of

completely novel animal forms in Cambrian strata in the sedimentary rock layers without transitional intermediate fossils—a finding contradictory to evolutionary theory.

82 Berlinski, David. *The Devil's Delusion: Atheism and Its Scientific Pretensions*. New York: Basic Books, 2009, pp. 191-2.

83. Wald, George. "Innovation and Biology." *Scientific American*, Vol. 199, September 1958, pp. 100.

84. Chen, Jean. "A Mom's Research (Part 5): A Deep Dive into Evolution." *The Epoch Times*, June 1, 2021 (updated June 29, 2021).

85. Dawkins, Richard. *The Greatest Show on Earth: The Evidence for Evolution*. Free Press, A Division of Simon & Schuster, Inc., 2009.

86. Sarfati, Jonathan. *The Greatest Hoax on Earth? Refuting Hawkins on Evolution*. Creation Book Publishers, 2014.

87. Gillispie C. C., Gratton-Guinness I., Fox R. *Pierre-Simon Laplace, 1749–1827: A Life in Exact Science*. Princetion, NJ: Princeton University Press, 1999.

88. 1 Corinthians 15:22

89. 1 Peter 4:12-13

90. 1 Peter 1:6-9

91. Doctrine & Covenants 130:20-21

92. Matthew 13:3-8, 19-23

93. John 15:1-6

94. Doctrine & Covenants 88:28-32

95. Hebrews 12:9

96. Romans 8:16-17

97. Luke 2:49; Moses 1:39 (Pearl of Great Price)
98. Psalms 82:6
99. Matthew 7:21; John 14:2; 2 Nephi 26:33
100. Doctrine & Covenants 101:37
101. John 17:3
102. Maxwell, Neal A. "Yet Thou Art There," *Ensign*, November 1987, pp. 30.
103. *Hymns of the Church of Jesus Christ of Latter-day Saints.* "Know This, That Every Soul Is Free." pp. 240.
104. Matthew 7:7-8
105. Matthew 6:32-33
106. Isaiah 55:8-9
107. Romans 14:11-12
108. Amos 3:7
109. Doctrine & Covenants 66:2
110. 2 Peter 1:4-11
111. Farrer, Austin. *Light on C.S. Lewis*, Jocelyn Gibb, ed. New York: Harcourt, Brace and World, Inc., 1966, pp. 26.
112. Slack, G., "What neo-creationists get right," *The Scientist*, 20 June 2008; the-scientist.com.
113. Moses 7:37, 40 (Pearl of Great Price)
114. Luke 17:12-19
115. Ephesians 2:12
116. Matthew 7:13-14
117. 2 Timothy 3:1-5, 7
118. Sill, Sterling W. *Leadership, Vol. 2*. Bookcraft, 1960, pp. 28.
119. Moroni 7:6, 9

120. Psalms 46:10

121. 1 Nephi 11:17

122. Proverbs 3:5; 2 Nephi 4:19

123. Matthew 25:1-13

124. Ephesians 4:14-15

125. Luke 15:7

126. 1 Corinthians 2:9

127. Sanford, J.C. *Genetic Entropy*, FMS Publications, 2014, pp. v.

128. Lewis, C.S., *The Weight of Glory.* Grand Rapids Michigan, The William B. Eerdmans Publishing Company, 1965, pp. 14-15.

129. Maxwell, Neal A. "A Summing Up." *Wherefore, Ye Must Press Forward.* Deseret Book Company, 1977, pp. 132-3.